W9-CFS-868

ADVANTAGE

Teacher's Guide for Assessment

- PORTFOLIO
- PROBLEM-SOLVING THINK ALONGS
- TEACHER AND STUDENT OBSERVATION CHECKLISTS
- MANAGEMENT FORMS

Grade 1

Harcourt Brace & Company

ndo • Atlanta • Austin • Boston • San Francisco • Chicago • Dallas • New York • Toronto • London

http://www.hbschool.com

Copyright © by Harcourt Brace & Company

All rights reserved. No part of this publication may be reproduced or
transmitted in any form or by any means, electronic or mechanical,
including photocopy, recording, or any information storage and retrieval
system.

Teachers using MATH ADVANTAGE may photocopy complete pages in
sufficient quantities for classroom use only and not for resale.

HARCOURT BRACE and Quill Design is a registered trademark of
Harcourt Brace & Company. MATH ADVANTAGE is a trademark of
Harcourt Brace & Company.

Printed in the United States of America

ISBN 0-15-311154-2

5 6 7 8 9 10 11   082   2004 2003 2002 2001

CONTENTS

▶ Overview
Assessment Model 1

▶ Description of Assessment Components
Daily Assessment 2
Formal Assessment 3
Performance Assessment 5
Portfolio Assessment 7
Student Self-Assessment 8

▶ Checklists, Rubrics and Surveys
Types Available and their Uses 9
Teacher-Observation Rubrics and Checklists
    Oral Presentation Scoring Rubric 10
    Project Scoring Rubric 11
    Cooperative Learning Checklist 12
Student Self-Observation Checklists and Surveys
    Group End-of-Project Checklist 13
    Individual End-of-Project Checklist 14
    Individual End-of-Chapter Checklist 15
    Attitudes Survey 16

▶ Interaction: A Key Factor in Instruction and Assessment
Teacher's Role 17
Children's Role 18
Family Involvement, Family Involvement Tips 19

▶ The Mathematics Portfolio
Building a Portfolio 20
Evaluating a Portfolio 21
Sharing the Portfolios, Benefits of Mathematics Portfolios 22
Portfolio Resource Sheets
    Portfolio Work Sample (Student) 23
    Portfolio Evaluation (Teacher) 24
    Portfolio Family Response (Parent) 25

► Assessing Problem Solving

Problem Solving Think-Along: Description...26

    Oral Response Form 27

    Written Response Form...28

► Management Suggestions and Forms

Description of Materials and Their Locations 29

    Test Answer Sheet

    25 Test-Taking Tips

    Grading Made Easy

    Individual Record Form

    Formal Assessment Class Record Form

    Performance Assessment Class Record Form

    Learning Goals 30

Overview

The assessment program in *Math Advantage* is both comprehensive and multidimensional. It allows learners a variety of opportunities to show what they know and can do, thus providing you with ongoing information about each student's understanding of mathematics. Equally important, the assessment program involves the student in self-assessment, offering you strategies for helping your students evaluate their own growth.

The *Math Advantage* assessment program is designed around the following assessment model.

**▶ *Math Advantage*
Assessment Model**

Key: **PE**–Pupil's Edition   **TE**–Teacher's Edition   **TGA**–Teacher's Guide for Assessment
     APK–Assessing Prior Knowledge   **TCM**–Test Copying Masters
     PA–Performance Assessment

Formal Assessment
- Inventory Test (TCM)
- Assessing Prior Knowledge (APK)
- Chapter Review/Test (PE/TE)
- Chapter Test (TCM)
- Multi-Chapter Review/Test (PE/TE)
- Multi-Chapter Test (TCM)
- Cumulative Reviews (PE/TE)
- Cumulative Test (TCM)

Performance Assessment
- Performance Assessments and Scoring Rubrics (PE/TE and PA)
- Interview/Task Tests and Evaluation (PA)
- Problem-Solving Think Along Response Sheets, Scoring Guides (TGA)
- Oral Presentation Scoring Rubric (TGA)
- Project Scoring Rubric (TGA)
- Cooperative Learning Checklist (TGA)

Daily Assessment
- Problem of the Day (TE)
- Mixed Review (PE/TE)
- Lesson Check (TE)

Student Self-Assessment
- Math Journal (TE)
- Group End-of-Project Checklist (TGA)
- Individual End-of-Project Checklist (TGA)
- End-of-Chapter Checklist (TGA)
- Attitudes Survey (TGA)

Portfolio Assessment
- Portfolio Work Sample, (TGA)
- Portfolio Evaluation (TGA)
- Family Response (TGA)

Daily Assessment

Daily Assessment is *assessment embedded in daily instruction.* Children are assessed as they learn and learn as they are assessed. First you observe and evaluate your children's work on an informal basis, and then you seek confirmation of those observations through other program assessments.

Math Advantage offers the following resources to support informal assessment on a daily basis.

Daily Assessment Measures

- **Problem of the Day,** in *Teacher's Edition* at beginning of lesson

- **Mixed Review,** in *Pupil's Edition* and *Teacher's Edition* following "Practice"

- **Lesson Check,** in *Teacher's Edition* in Wrap-Up at end of lesson

---

Options and Suggestions

✓ **Problem of the Day** kicks off the lesson with a problem that is relevant both to lesson content and the children's world. Its purpose is to spark student thinking about the lesson topic and provide insights about children's ability to solve a problem related to it. Class discussion may yield clues about children's readiness to learn a concept or skill emphasized in the lesson.

✓ "Child-watching" is a natural and important part of daily assessment. As you teach the lesson, you may want to make a mental note of important observations (or devise a simple way to jot them down and keep them for future reference).

✓ You may want to use **Mixed Review,** an assessment feature in the *Pupil's Edition,* to assess skills taught in previous lessons. The **Spiral Review** in the *Teacher's Edition* is another resource for reviewing.

✓ To assess lesson content, use **Wrap Up and Assess** in the *Teacher's Edition* at the end of each lesson. This feature includes two brief assessments.

    —a question to probe children's grasp of the main lesson concept

    —a quick lesson check on children's mastery of lesson skills

✓ Depending on what you learn from children's responses to end-of-lesson assessments, you may wish to use **Problem Solving, Reteach, Practice,** or **Enrichment** copying masters before starting the next lesson.

Formal Assessment

Formal Assessment in *Math Advantage* consists of a series of reviews and tests that assess how well children understand concepts, perform skills, and solve problems related to program content. Information from these measures (along with information from other kinds of assessment) is needed to evaluate a child's achievement and to determine grades. Moreover, analysis of results can help determine whether additional practice or reteaching is needed.

Formal Assessment in *Math Advantage* includes the following measures.

Formal Assessment Measures

- **Inventory Test,** in *Test Copying Masters*

- **Assessing Prior Knowledge** screening worksheet, in *Assessing Prior Knowledge*; shown in *Teacher's Edition* on Chapter Overview page

- **Chapter Review/Test** in *Pupil's Edition* and *Teacher's Edition*

- **Chapter Test** in *Test Copying Masters*, reduced in *Teacher's Edition*

- **Multi-Chapter Review** in *Pupil's Edition* and *Teacher's Edition*

- **Multi-Chapter Test** in *Test Copying Masters*, reduced in *Teacher's Edition*

- **Cumulative Reviews** in *Pupil's Edition* and *Teacher's Edition*

- **Cumulative Test** in *Test Copying Masters*, reduced in *Teacher's Edition*

Options and Suggestions

✓ Formal tests are available in two formats, multiple-choice and free-response. At times, you may choose to use the multiple-choice test because its format helps prepare children for the standardized achievement tests. At other times, you may want to use the free-response test because it gives you diagnostic information about each child—information you need to select from the various practice or reteaching options the program offers.

✓ The Inventory Test, available on copying masters, is a formal assessment tool that assesses how well children have mastered the objectives of the previous grade level. Test results provide information about the kinds of review your children may need to be successful in mathematics at their new grade level. You may wish to use it at the beginning of the school year or when a new child arrives.

✓ Before beginning a chapter, you may wish to give a simple screening worksheet to find out whether your children have the skills necessary for success in that chapter.

You can find such screening worksheets in *Assessing Prior Knowledge,* a separate assessment book. A **Review Suggestions** chart accompanies each answer key for the worksheet. It gives suggestions for building the concepts and skills that results may show children lack. The chart suggests specific intervention lessons or Practice Activities that can help develop each concept or skill.

You may wish to use a pretest and posttest with the chapter. Use either the Form A Test, multiple-choice, or the Form B Test, free-response, as a pretest and the alternate form as a posttest.

✓ **Chapter Review/Test** is an assessment feature in the *Pupil's Edition.* Use it at the end of a chapter to reinforce learning and determine whether there is a need for more instruction or practice. Discussion of responses can help correct lingering misconceptions before children take the Chapter Test.

✓ The **Chapter Test,** in *Test Copying Masters,* is available in two formats— multiple-choice, Form A, and free-response, Form B. Both test the same content. The two forms can be used to provide a pretest and a posttest or as two forms of the posttest.

✓ Children can record their answers on the test sheet. However, for multiple-choice tests that have no more than 50 items, you may choose to have them use the *Answer Sheet,* similar to the "bubble form" used for standardized tests, that is located in *Test Copying Masters.*

✓ **Multi-Chapter Review/Test** appears in the Pupil's Edition, at the end of every content cluster. This feature presents problems that relate to content from the group of chapters and gives examples to help children solve them. Teacher guidance is recommended.

✓ The **Multi-Chapter Test,** available on copying masters, deals with the same content as the review. It assesses understanding of key ideas and ability to demonstrate skills emphasized in the content cluster.

✓ The **Cumulative Reviews** are similar to the **Multi-Chapter Tests,** but the scope broadens from a few chapters to all chapters up to and including the current one. Discussion of children's responses can help prepare them for the **Cumulative Test,** which is available on copying masters.

✓ The **Answer Key,** in *Test Copying Masters,* provides a reduced replication of the test with answers.

✓ Several test record forms are available for Formal Assessment; each serves a different purpose. These forms are listed at the end of this booklet in the section titled "Management Suggestions and Forms."

✓ Look for patterns in test results. They can signal adjustments in instruction or assessment that can help meet group and individual needs.

Performance Assessment

In the past, children's grades in math were based almost solely on traditional test scores. Teachers today have come to realize that the multiple-choice format of these tests, while useful and efficient, cannot provide a complete picture of children's growth. Standardized tests may show what students know, but they are not designed to show how they *think and do things*—an essential aspect of math literacy. Performance Assessment, together with other types of assessments, can supply the missing information and balance your assessment program. Perfomance Assessments, in particular, help reveal the thinking strategies children use to work through a problem, and children usually enjoy them more than standardized tests.

Math Advantage offers the following assessment measures, scoring instruments, and teacher observation checklists for evaluating a child's performance.

Performance Assessment Measures

- **Performance Assessments** in *Pupil's Edition* with **Scoring Rubrics** in *Teacher's Edition*

- **Quarterly, Extended Performance Assessments** and **Scoring Rubrics,** in *Performance Assessment*

- **Interview/Task Tests and Evaluation,** in *Performance Assessment*

- **Problem-Solving Think Alongs, Response Sheets, and Scoring Guides,** in the pages that follow

- **Oral Presentation Scoring Rubric,** in the pages that follow

- **Project Scoring Rubric,** in the pages that follow

- **Cooperative Learning Checklist,** in the pages that follow

---

Options and Suggestions

✓ Before children begin a performance task, discuss how they will be evaluated. You may choose to develop scoring rubrics with your children or to use those offered in the program.

✓ If you wish, interact with children as they complete a task. A question from you that encourages reflection may be all it takes to help a puzzled child proceed. You should motivate, guide, and challenge children to produce their best work—without actually doing the work for them.

✓ Within the Pupil's Edition are periodic Assessment Checkpoints. These sections always include one page of *Performance Assessment*. The Teacher's Edition has a matching scoring rubric for the tasks and problems.

✓ The assessment program at each level includes four **Performance Assessments,** each of which can be completed in one or two class sessions. These assessments can help you assess children's ability to use what they have learned to solve everyday problems. Each assessment focuses on a theme and includes four separate tasks. You may want to use this type of assessment quarterly or at the end of grading periods to help children prepare for district or state performance tests. The four assessments, along with a scoring rubric and student work samples, are available in *Performance Assessment Book.*

✓ The **Interview/Task Test** is a "one-on-one" test that facilitates teacher-child interaction. As such, it is especially useful for assessing children who perform poorly on standardized tests. The evaluation criteria for each task helps you pinpoint errors in the child's thinking.

✓ The **Problem-Solving Think Along** is a self-questioning performance assessment that is designed around the problem-solving process used in *Math Advantage.* It is available in two forms. Either form can be used to assess performance as children work through the process. Each form has its own scoring guide.

> **Oral Response Form**—a handy interview instrument to assess student performance in problem solving (students verbalize their thinking as they work through the process)

> **Written Response Form**—a form that individuals or groups can use to record the process they use to solve a problem

✓ The observation checklists, listed below, provide a way for you to evaluate three important classroom activities. Each checklist offers criteria for evaluation.

> **Oral Presentation Scoring Rubric**—for evaluating an individual or group presentation that may be lengthy, such as describing a project, or brief, such as demonstrating a way to solve a problem; also a handy self-checking guide for children to use during the planning stages of a presentation.

> **Project Scoring Rubric**—for evaluating an individual or group project. A project is an open-ended, problem-solving activity that may involve activities such as gathering data, constructing a data table or graph, writing a report, or building a model.

> **Cooperative Learning Checklist**—for evaluating a child's behavior as he or she works in a group; also for guiding a discussion of ways in which a child can become a more effective group member.

▶ Portfolio Assessment

A portfolio is a collection of child-selected and teacher-selected work samples that represent the individual's accomplishments and growth over a period of time. The main purpose of a portfolio in mathematics is to provide both the teacher and the child with a concise, yet comprehensive, picture of the student's progress in the subject.

Support materials for building and evaluating portfolios are listed below.

Portfolio Support Materials

- *A Guide to My Math Portfolio,* in the pages that follow

- *Portfolio Evaluation*, in the pages that follow

- *Family Response,* in the pages that follow

Options and Suggestions

✓ Explain to children that the purpose of their Math Portfolio is to show samples of work that demonstrate their growth in mathematics. Point out that the best sample is not necessarily the neatest paper or the one with the highest score. Discuss the kinds of selections that might best show evidence of growth in what a child knows and can do. Activities that may produce especially useful work samples are identified in the chapter.

✓ Establish a basic plan that shows how many child-selected and teacher-selected work samples will go into the portfolio during a certain period of time and when they should be selected.

✓ Ask children to make a list of their work samples on **A Guide to My Math Portfolio.** Have them tell what they have learned. This reflective activity builds self-evaluation and decision-making skills and encourages children to organize their portfolios in a thoughtful manner.

✓ Discuss portfolios at regular intervals. Compare the child's current portfolio to his or her previous one, rather than to those of others. Record evidence of growth on the **Portfolio Evaluation** sheet. You may wish to list some things the student might do to improve his or her next evaluation. Attach the completed evaluation sheet to the portfolio. Use it to help make conferences with children and their parents a positive experience.

✓ If children take their portfolios home to share with family members, you may want to include **Family Response,** a home-involvement letter, that requests parental review of the child's portfolio. At the bottom of the sheet is a place for a family member's comments. The Family Response Sheet should be returned to school with the portfolio.

Student Self-Assessment

Research shows that self-assessment can have significant positive effects on a child's learning. To achieve these effects, children must be challenged to reflect on their work and to monitor, analyze, and control their learning. Their ability to evaluate their behaviors and to monitor them grows with their experience in self-assessment.

Math Advantage offers the following self-assessment tools for your use.

Self-Assessment Resources

- **Math Journal,** ideas for journal writing in Teacher's Edition

- **Group End-of Project Checklist,** in the pages that follow

- **Individual End-of-Project Checklist,** in the pages that follow

- **End-of Chapter Checklist,** in the pages that follow

- **Attitudes Survey,** in the pages that follow

Options and Suggestions

✓ The **Math Journal** is a collection of a child's writings that may communicate feelings, ideas, and explanations as well as responses to open-ended problems. It is an important evaluation tool in math, even though it is not graded. Use the journal to gain important insights about a child's growth that you cannot obtain from other assessments. Look for journal icons in your Teacher's Edition for suggested journal-writing activities.

✓ The **Group End-of Project Checklist,** titled "How Well Did Our Group Do?" is designed to assess and build group self-assessment skills, while the Individual End-of-Project Checklist, titled "How Well Did I Work in My Group?" helps the child evaluate his or her own behavior in the group.

✓ The **Individual End-of-Chapter Checklist,** titled "How Did I Do?" leads children to reflect on what they have learned and how they learned it. Use it to help them learn more about their own capabilities and develop confidence.

✓ The **Attitudes Survey,** titled "How I Feel About Math," focuses on children's attitudes about math. Use it in regular intervals to monitor changes in individual and group attitudes.

✓ Discuss directions for completing each checklist you use and tell children there are no "right" answers to the items. Discuss reasons for various responses.

✓ Give students time to elaborate on their responses in student conferences.

▶ Checklists, Rubrics, and Surveys

Types Available and Their Uses

Two types of assessment tools are offered in this section: classroom observation measures to help teachers evaluate children's performance, and self-assessment tools to help children evaluate their own efforts. Scoring rubrics and teacher-observation checklists are available for assessing children's oral presentations, projects, and participation in cooperative learning groups. Self-evaluation checklists give children a chance to reflect upon their work as individuals and as members of a group. They also lead them to think about what they are learning and their attitudes about math.

These checklists give you information about children's confidence, flexibility, willingness to persevere, interest, curiosity, inventiveness, inclination to monitor and reflect on their own thinking and doing, and appreciation of the role of mathematics in our culture.

Child's Name _____ Date _____

Oral Presentation Scoring Rubric

Check the indicators that describe the child's presentation. Use the location of the check marks to determine the individual's or group's overall score.

3 Point Score Indicators: The presentation

_____ shows evidence of extensive research/reflection.

_____ demonstrates thorough understanding of content.

_____ is exceptionally clear and effective.

_____ exhibits outstanding insight/creativity.

_____ is of high interest to the audience.

2 Point Score Indicators: The presentation

_____ shows evidence of adequate research/reflection.

_____ demonstrates acceptable understanding of content.

_____ overall is clear and effective.

_____ shows reasonable insight/creativity.

_____ is of general interest to the audience.

1 Point Score Indicators: The presentation

_____ shows evidence of limited research/reflection.

_____ demonstrates partial understanding of content.

_____ is clear in some parts but not in others.

_____ shows limited insight/creativity.

_____ is of some interest to the audience.

0 Point Score Indicators: The presentation

_____ shows little or no evidence of research/reflection.

_____ demonstrates poor understanding of content.

_____ for the most part is unclear and ineffective.

_____ does not show insight/creativity.

_____ is of little interest to the audience.

Overall score for the presentation. _____

Comments: _____

_____

_____

Project Scoring Rubric

Check the indicators that describe a child's or group's performance on a project. Use the location of the checkmarks to help determine the individual's or group's overall score.

3 Point Score Indicators: The group

_____ makes outstanding use of resources.

_____ shows thorough understanding of content.

_____ demonstrates outstanding grasp of mathematics skills.

_____ displays strong decision-making/problem-solving skills.

_____ exhibits exceptional insight/creativity.

_____ communicates ideas clearly and effectively.

2 Point Score Indicators: The group

_____ makes good use of resources.

_____ shows adequate understanding of content.

_____ demonstrates good grasp of mathematics skills.

_____ displays adequate decision-making/problem-solving skills.

_____ exhibits reasonable insight/creativity.

_____ communicates most ideas clearly and effectively.

1 Point Score Indicators: The group

_____ makes limited use of resources.

_____ shows partial understanding of content.

_____ demonstrates limited grasp of mathematics skills.

_____ displays weak decision-making/problem-solving skills.

_____ exhibits limited insight/creativity.

_____ communicates some ideas clearly and effectively.

0 Point Score Indicators: The group

_____ makes little or no use of resources.

_____ fails to show understanding of content.

_____ demonstrates little or no grasp of mathematics skills.

_____ does not display decision-making/problem-solving skills.

_____ does not exhibit insight/creativity.

_____ has difficulty communicating ideas clearly and effectively.

Overall score for the project. _____

Comments: _____

_____

_____

Child's Name _____ Date _____

Cooperative Learning Checklist

Ring the response that best describes the child's behavior.

Never   Behavior is not observable.

Sometimes Behavior is sometimes, but not always, observable when appropriate.

Always   Behavior is observable throughout the activity or whenever appropriate.

<u>The child</u>

• is actively involved in the activity.	Never	Sometimes	Always
• shares materials with others.	Never	Sometimes	Always
• helps others in the group.	Never	Sometimes	Always
• seeks the teacher's help when all group members need help.	Never	Sometimes	Always
• fulfills his or her assigned role in the group.	Never	Sometimes	Always
• dominates the activity of the group.	Never	Sometimes	Always
• shares ideas with others.	Never	Sometimes	Always
• tolerates different views within the group about how to solve problems.	Never	Sometimes	Always

> **Use this checklist to discuss each child's successful cooperative learning experiences and ways in which he or she can become a more effective group member.**

Group _____

Date _____

How Well Did Our Group Do?

Members _____

_____

This picture or writing tells about our project.

```
┌─────────────────────────────────────────┐
│                                         │
│                                         │
│                                         │
│                                         │
│                                         │
│                                         │
└─────────────────────────────────────────┘
```

1. How well did we share ideas?

2. How well did we work together?

3. How well did we solve problems?

4. How well did we put things away?

5. How well did our project turn out?

Name _____

Date _____

How Well Did I Do?

_____

Our project was about _____.

1. How hard did I work on the project?   😊   ☹️

2. How well did I work with others?   😊   ☹️

3. How well did I share my ideas?   😊   ☹️

4. How well did I listen to others?   😊   ☹️

5. How well did I understand the problem?   😊   ☹️

6. How much did I learn?   😊   ☹️

What I liked most about the project was

```



```

Name _____

Date _____

Think Back

_____
_ _ _ _ _ _ _ _ _ _ _ _ _ _ _ _ _

The chapter was about _____.

1. I liked this chapter.   Yes   ?   No

2. I understood the ideas.   Yes   ?   No

3. I worked hard to learn in class.   Yes   ?   No

4. I asked questions if I did not
   understand something.   Yes   ?   No

5. I did a good job of solving problems
   in this chapter.   Yes   ?   No

6. I practiced solving problems
   outside of school.   Yes   ?   No

7. I learned a lot from this chapter.   Yes   ?   No

8. This is one thing I learned.

How I Feel About Math

1. I like solving problems in math.

2. I like telling how I solved a problem.

3. I like working alone in math.

4. I like working with others.

5. I think math is easy.

6. Math is one of my best subjects.

7. Math activities are fun.

This is the activity I like the best.

▶ Interaction:
A Key Factor in Instruction and Assessment

We believe that

- children develop mathematical power when they think consistently, communicate, and reason; draw on mathematical ideas; use tools and techniques effectively; reflect on their work; and make revisions based on that reflection.

- children learn to value diversity and to understand and appreciate the viewpoints of others when they interact with their peers.

- children take responsibility for their learning as they ask questions, formulate problems, and make decisions about what to do.

Math Advantage is a program that is centered on interaction. Children are encouraged to share ideas about what they have learned and to listen and learn what is important about the ideas being shared by others. As children work together, they agree and disagree and often learn mathematics as a result of student-student and student-teacher interaction.

Teacher's Role

The role of the teacher in mathematics is that of a facilitator posing engaging problems and tasks to children and then challenging the children to work together to solve the problems and share strategies. You can promote interaction among children in the classroom, encouraging them to explain their thinking and reasoning. Through observing and listening to children, you can guide and evaluate them in ways that are best for them.

Questioning techniques are critical strategies to ensure that the children's ideas and conclusions are based on their explorations and information. After asking a question, use **wait time** to allow the child more time to answer. If a child is unsure of an answer, restate the question or rephrase it, using different words that might help the child determine the answer. In this way you can encourage open discussion and exploration.

In *Math Advantage,* children are presented with a problem or task in each lesson or project. It is important that you provide time for children to explore and to communicate with each other. You observe them working together, and based on the observations, you select questions provided that are appropriate to the children's progress on the problem or task. It is through observation and interaction that you are able to assess a child's depth of understanding and to guide the child to the next step. You evaluate each child by observing and questioning as the child explains his or her thinking. The Portfolio Conference provides a time for you and the child to review the child's work and discuss his or her growth in mathematics.

Children's Role

Primary-age children can take active roles in their own learning. They should be encouraged to choose many kinds of manipulatives and tools when working to solve problems and tasks. Through multiple representations of the same problem, children learn to generalize and draw conclusions. They share ideas with classmates and discuss problems encountered, discoveries made, and strategies used. Through the interaction, mathematics is learned. By representing a problem in several ways, and by providing children opportunities to work on open-ended tasks and to choose their own manipulatives, you improve the quality of their interaction.

One of the most important and difficult things for a child to do is to be a good listener. Children love to express their thoughts and ideas to others, but they need to be taught to listen while others express their thoughts and ideas. Children need to understand that what others say is also important and that they might learn from them. In working together to solve problems, children should learn to question what they are doing and what they have done to test their ideas. When children share their conclusions and knowledge with others, they expand their own understanding of mathematics. By doing so, they build self-confidence and the ability to evaluate their own work.

In *Math Advantage,* children work together to construct their learning. They are encouraged to openly share and discuss ideas and strategies. Through solving problems and tasks, children develop ideas and draw conclusions about big ideas in math. Children are encouraged to test their conclusions and revise them if they feel they are incorrect or if there is a better way to solve the problem. At the end of each lesson, children express what they learned from each experience.

The classroom environment should be one in which children feel comfortable sharing discoveries and ideas with others. Children should be good listeners and receptive to others' thoughts and ideas. As children talk together about mathematics, they clarify and verify their thinking and develop deeper understanding of important concepts and ideas.

Family Involvement

Family involvement in school activities helps children learn more effectively. You should keep the family informed of the content being covered as well as the growth the child is making in mathematics. Help family members understand the methods being used to evaluate the child's progress and why they are important. Family members should be encouraged to review or extend what the child is learning and experiencing at school.

Informal features designed to encourage family involvement are found throughout the program. They include the following:

- A four-page brochure that introduces family members to the *Math Advantge* program at the beginning of the school year.

- A letter describing what children will learn and suggesting activities that can be done at home to support learning is located at the beginning of each group of chapter content.

- Home Notes are located on each student activity page to keep the family informed about the mathematics concepts children are learning.

- Student Journals containing a written record of children's work and their thoughts about their work can be sent home periodically to be shared with family members.

- An Assessment Portfolio with samples of the child's work, the results of the Performance Assessment Tasks, and the written assessment can be sent home periodically. In this way, the teacher presents family members with a clear picture of their child's progress. Family members should be encouraged to respond by writing in the portfolio to express the growth they see in their child's performance.

Family Involvement Tips

- Schedule periodic conferences with family members and the child to share and discuss the contents of the child's portfolio and to discuss goals for the child's mathematical growth.

- Invite family members to visit the classroom to observe or participate while the children are working on a problem or task.

- Invite children and their family members to participate in a "Family Math Night," during which they work together, using a variety of manipulatives to solve problems or tasks similar to those worked on during class time.

- Encourage family members to collect or make materials needed for math class.

- Encourage family members to help the child create a game or puzzle that uses math skills.

The Mathematics Portfolio

The portfolio is a collection of each child's work gathered over an extended period of time. A portfolio illustrates the growth, talents, achievements, and reflections of the mathematics learner and provides a means for the teacher to assess the child's performance.

An effective portfolio will

- include items collected over the entire school year.

- show the "big picture"—providing a broad understanding of a child's mathematics language and feelings through words, diagrams, checklists, and so on.

- give children opportunities to collaborate with their peers.

- give children a chance to experience success, to develop pride in their work, and to develop positive attitudes toward mathematics.

Building a Portfolio

There are many opportunities to collect students' work throughout the year as you use *Math Advantage.* A list of suggested portfolio items for each chapter is given in Chapter Overview in your teacher's edition. Give children the opportunity to select some of their work to be included in the portfolio.

To begin:

- Provide a file folder for each child with the child's name clearly marked on the tab or folder.

- Explain to children that throughout the year they will save some of their work in the folder. Sometimes it will be their individual work; sometimes it will be group reports and projects, or completed checklists.

- Assign a fun activity to the entire class that can be placed into the portfolio by all children. EXAMPLE: Ask children to draw a map of the school playground in which they use at least three different shapes.

- Comment positively on the maps and reinforce the process. Have children place their maps into their portfolios.

Evaluating a Portfolio

The ultimate purpose of assessment is to enable children to evaluate themselves. Portfolios have the potential to create authentic portraits of what children learn and offer an alternative means for documenting growth, change, and risk-taking in mathematics learning. Evaluating their own growth in mathematics will be a new experience for most children. The following points made with regular portfolio evaluation will encourage growth in self-evaluation.

- Discuss the contents of the portfolio with each child as you examine it at regular intervals during the school year.

- Examine each portfolio on the basis of the growth the child has made rather than in comparison with other portfolios.

- Ask the child questions as you examine the portfolio.

- Note statements that express how the child feels.

- Point out the strengths and weaknesses in the work.

- Encourage and reward the child by emphasizing the growth, the original thinking, and the completion of tasks you see.

- Reinforce and adjust instruction of the broad goals you want to accomplish as you evaluate the portfolios.

What To Look For

Growth in mathematics is shown by:

- the non-standard responses that children make.

- the ways children solve a problem or attempt to solve a problem. Note the strategies they used and the reasons they succeeded or became confused.

- the ways children communicate their understanding of math problems. Do they use words, pictures, or abstract algorithms?

- unique solutions or ways of thinking.

Placing comments such as the following on children's work can have instructional benefits.

"I like the way you did this."
"Is this another way to solve the problem?"
"Can you think of another problem like this one?"

Sharing the Portoflios

- Examine the portfolio with family members to share concrete examples of the work the child is doing. Emphasize the growth you see as well as the expectations you have.

- Examine portfolios with your children to emphasize their experiences of success and to develop pride in their work and positive attitudes toward mathematics.

- Examine portfolios with your supervisor to share the growth your children have made and to show the ways you have developed the curriculum objectives.

The Benefits of Mathematics Portfolios

Portfolios can be the basis for informed change in mathematics classrooms because they:

- send positive messages to children about successful processes rather than end results.

- give you better insights as to how children understand and work problems.

- focus on monitoring the development of reasoning skills.

- help children become responsible for their own learning.

- promote teacher-child interaction.

- focus on the child, rather than on the assignment.

- focus on the development of conceptual understandings, rather than applications of skills and procedures.

Name _____

Date _____

See What I Can Do

I chose this work for my portfolio.

This picture or writing tells about it.

I chose this work because it shows

_____

_____

_____

_____

Name _____

Date _____

Evaluating Performance	Evidence and Comments
1. What mathematical understandings are demonstrated?	_____ _____ _____
2. What skills are demonstrated?	_____ _____ _____
3. What approaches to problem solving and critical thinking are evident?	_____ _____ _____
4. What work habits and attitudes are demonstrated?	_____ _____ _____

Summary of Portfolio Assessment

For This Review			Since Last Review		
Excellent	Good	Fair	Improving	About the Same	Not as Good

Date _____

Dear Family,

Here are samples of math work that your child and I have chosen for portfolio assessment. Please ask your child to explain what he or she has done. Then write a short note to your child in the space below, telling your thoughts about what you have seen. Please have your child bring the portfolio, with your note, back to school.

Sincerely,

_____
(Teacher)

- -

Response to Portfolio:

Dear _____ ,

_____
(Family member)

Assessing Problem Solving

Assessing a child's ability to solve problems in *Math Advantage* involves more than checking the answer. It involves looking at how the child processes information and how he or she works at solving problems. The heuristic used in the program's Problem-Solving Think Along—Understand, Plan, Solve, and Look Back—guides the child's thinking process and provides a structure within which he or she can work toward a solution. Evaluating the child's progress through the parts of the heuristic can help you assess areas of strength and weakness in solving problems.

These instruments may be used to assess children's problem-solving abilities.

Problem-Solving Think Along

Oral Response Form (See page 27.)

This form can be used by a child or group as a self-questioning instrument or as a guide for working through a problem. It can also be an interview instrument the teacher can use to assess children's problem-solving skills and can be used with the Interview/Task Test.

Problem-Solving Think Along

Written Response Form (See page 28.)

This form provides a recording sheet for a child or group to record their responses as they work through each section of the heuristic. This Written Response Form is also used to record children's responses to the problem-solving items in the Interview/Task Test for each chapter.

Problem-Solving Think Along:
Oral Response Form

Understand and Plan

1. Explain the problem in your own words.

2. Where will you get the information?

3. What materials will you need?

4. What job will each person do?

Solve

5. What did I find out?

6. How can I record the information?

Look Back

7. Read the problem again. Think about what you have done.
Do you want to change anything? Explain.

8. Is your answer reasonable? Explain.

9. What is another way to solve the problem?

Understand

1. Tell the problem in your own words.

_____

_____

_____

2. What do you want to find out?

_____

_____

_____

Plan

3. How will you solve the problem?

_____

_____

_____

Solve

4. Show how you solved the problem.

_____

_____

_____

Look Back

5. How can you check your answer?

_____

_____

_____

Management Suggestions and Forms

Description of Materials and Their Locations

- **Test Answer Sheet,** in *Test Copying Masters*
  This copying master is an individual recording sheet for up to 50 items on a multiple-choice or standardized-format test.

- **25 Test-Taking Tips,** in *Test Copying Masters*
  Help your students do their best on standardized-format tests by talking over these these "do's" and "don'ts" on preparing for and taking a test.

- **Grading Made Easy,** in *Test Copying Masters*
  This percent converter can be used for all quizzes and tests. The percents given are based on all problems having equal value. Percents are rounded to the nearest whole percent, giving the benefit of 0.5 percent.

- **Individual Record Form,** in *Test Copying Masters*
  This form provides a place to enter a child's scores on all formal tests and to indicate the objectives he or she has met. Criterion scores for each learning goal are given. A list of review options is also included. The options include activities in the Pupil's Edition, Teacher's Edition, Workbooks, and other places where you will find activities that you can assign to the child who is in need of additional practice.

- **Formal Assessment Class Record Form,** in *Test Copying Masters*
  This form makes it possible to record the test scores of an entire class on a single form. Criterion scores are listed for each of the tests.

- **Performance Assessment Class Record Form,** in *Performance Assessment*
  Use this record form to check the skills your students demonstrate on the Interview Task/Tests and on other performance assessments.

- **Learning Goals,** *Teacher's Guide for Assessment*
  The learning goals for this grade level are on the pages that follow. You may use this checklist to coordinate the program's learning goals with district or state objectives.

Learning Goals

Child's Name _____ Teacher _____

CHAPTER 1

Test Score:	Criteria Form A Form B 8/12 _____ _____	Needs More Work	Accomplished
1-A.1 ☐ ☐	To use concrete materials to model addition story problems		
1-A.2 ☐ ☐	To add on 1 or 2 to find sums to 6		
1-A.3 ☐ ☐	To use pictures to find sums		
1-A.4 ☐ ☐	To write and solve addition sentences to represent addition story problems		

CHAPTER 2

Test Score:	Criteria Form A Form B 8/12 _____ _____	Needs More Work	Accomplished
2-A.1 ☐ ☐	To use concrete materials to model subtraction story problems		
2-A.2 ☐ ☐	To subtract 1 or 2 to find differences from 6		
2-A.3 ☐ ☐	To write and solve subtraction sentences to represent pictures		
2-A.4 ☐ ☐	To use the strategy *make a model* to solve addition and subtraction story problems		

CHAPTER 3

Test Score:	Criteria Form A Form B 8/12 _____ _____	Needs More Work	Accomplished
3-A.1 ☐ ☐	To use counters to understand the Commutative Property of Addition		
3-A.2 ☐ ☐	To identify combinations of addends with sums to 10		
3-A.3 ☐ ☐	To add basic facts to 10 in vertical and horizontal formats		
3-A.4 ☐ ☐	To use the strategy *make a model* to solve addition problems with money		

CHAPTER 4

Test Score:	Criteria Form A Form B 8/12 _____ _____	Needs More Work	Accomplished
4-A.1 ☐ ☐	To use the strategy *counting on* to add		
4-A.2 ☐ ☐	To use counters to show doubles and write the fact		
4-A.3 ☐ ☐	To find sums for addition facts to 10		
4-A.4 ☐ ☐	To use the strategy *drawing a picture* to solve problems		

Use the boxes ☐ ☐ to indicate cross-references to local, district, or statewide benchmarks.

Learning Goals

Child's Name _____ Teacher _____

CHAPTER 5

Criteria Form A Form B 8/12 _____ _____	Needs More Work	Accom-plished
To identify combinations of ways to subtract from numbers 10 or less		
To subtract in horizontal and vertical format		
To identify fact families with facts to 10		
To use subtraction to compare two groups of 10 or less		

CHAPTER 6

Criteria Form A Form B 8/12 _____ _____	Needs More Work	Accom-plished
To use the *counting back* strategy to subtract		
To subtract zero and subtract to find a difference of zero		
To find sums and differences for addition and subtraction facts to 10		
To use the strategy *drawing a picture* to solve problems		

CHAPTER 7

Test Score: 7/10	Criteria Form A Form B _____ _____	Needs More Work	Accom-plished
7-A.1 ☐ ☐	To identify solid figures		
7-A.2 ☐ ☐	To sort solid figures by attributes		
7-A.3 ☐ ☐	To use the strategy *make a model* to identify how many cubes are used to build a figure		

CHAPTER 8

Test Score: 8/12	Criteria Form A Form B _____ _____	Needs More Work	Accom-plished
8-A.1 ☐ ☐	To identify plane figures		
8-A.2 ☐ ☐	To sort plane figures by the number of sides and corners		
8-A.3 ☐ ☐	To identify figures that have the same size and shape		
8-A.4 ☐ ☐	To identify a line of sym-metry in plane figures		

Learning Goals

Child's Name _____ Teacher _____

CHAPTER 9

Test Score:	Criteria Form A Form B 7/10 _____ _____	Needs More Work	Accom-plished
9-A.1 ☐ ☐	To identify plane figures as open or closed		
9-A.2 ☐ ☐	To classify objects by position		
9-A.3 ☐ ☐	To locate positions on a grid		

CHAPTER 10

Test Score:	Criteria Form A Form B 7/10 _____ _____	Needs More Work	Accom-plished
10-A.1 ☐ ☐	To identify, reproduce, and extend patterns		
10-A.2 ☐ ☐	To create patterns		
10-A.3 ☐ ☐	To analyze and correct patterns		

CHAPTER 11

Test Score:	Criteria Form A Form B 12/16 _____ _____	Needs More Work	Acc plis
11-A.1 ☐ ☐	To use strategies such as counting on and doubles to find sums to 12		
11-A.2 ☐ ☐	To add three addends with sums through 12		
11-A.3 ☐ ☐	To solve addition story problems by acting them out and writing addition sentences to represent the problems		

CHAPTER 12

Test Score:	Criteria Form A Form B 8/12 _____ _____	Needs More Work	Acc plis
12-A.1 ☐ ☐	To use mental math strategies such as counting back, relating addition and subtraction, and fact families to find differences to 12		
12-A.2 ☐ ☐	To compare groups of objects to find the differences between them		
12-A.3 ☐ ☐	To solve addition and subtraction story problems by writing a number sentence		

Learning Goals

Child's Name _____ Teacher _____

APTER 13

t re:	Criteria Form A Form B 8/12 ____ ____	Needs More Work	Accom- plished
A.1 ⊐	To count groups of tens, identify and write the number		
A.2 ⊐	To count groups of tens and ones to 100, identify and write the number		
A.3 ⊐	To use 10 as a benchmark to estimate a quantity as more than or fewer than 10		

APTER 14

t re:	Criteria Form A Form B 8/12 ____ ____	Needs More Work	Accom- plished
A.1 ⊐	To identify ordinal numbers from *first* through *twelfth*		
A.2 ⊐	To compare two numbers and identify which number is greater or less		
A.3 ⊐	To identify numbers that come before, after, or between other numbers		
A.4 ⊐	To order numbers (less than 100) from least to greatest or from greatest to least		

CHAPTER 15

Test Score:	Criteria Form A Form B 7/10 ____ ____	Needs More Work	Accom- plished
15-A.1 ☐ ☐	To count by twos, fives, and tens to 100		
15-A.2 ☐ ☐	To identify a number as even or odd		

CHAPTER 16

Test Score:	Criteria Form A Form B 7/10 ____ ____	Needs More Work	Accom- plished
16-A.1 ☐ ☐	To count groups of pennies and groups of nickels and give the value		
16-A.2 ☐ ☐	To count groups of dimes and give the value		
16-A.3 ☐ ☐	To count combinations of nickels and pennies and give the value		
16-A.4 ☐ ☐	To count combinations of dimes and pennies and give the value		
16-A.5 ☐ ☐	To use the strategy *make a model* to identify coins to use to buy an object		

Learning Goals

Child's Name _____ Teacher _____

CHAPTER 17

Test Score:	Criteria Form A Form B 7/10 _____ _____	Needs More Work	Accom-plished
17-A.1 ☐ ☐	To trade pennies for nickels and dimes		
17-A.2 ☐ ☐	To use combinations of coins to show a given amount using the fewest coins		
17-A.3 ☐ ☐	To identify what coins are needed to purchase an item		
17-A.4 ☐ ☐	To make combinations of pennies, nickels, and dimes that represent the value of a quarter		

CHAPTER 18

Test Score:	Criteria Form A Form B 7/10 _____ _____	Needs More Work	Accom-plished
18-A.1 ☐ ☐	To read a calendar		
18-A.2 ☐ ☐	To sequence events		
18-A.3 ☐ ☐	To estimate which of two tasks will take more time		

CHAPTER 19

Test Score:	Criteria Form A Form B 7/10 _____ _____	Needs More Work	Acc plis
19-A.1 ☐ ☐	To tell time to the hour and half hour		
19-A.2 ☐ ☐	To estimate the time needed to do a task as more or less than one minute		

CHAPTER 20

Test Score:	Criteria Form A Form B 7/10 _____ _____	Needs More Work	Acc plis
20-A.1 ☐ ☐	To use nonstandard units to measure length		
20-A.2 ☐ ☐	To measure the length of objects in inches		
20-A.3 ☐ ☐	To measure the length of objects in centimeters		

Learning Goals

Child's Name _____   Teacher _____

CHAPTER 21

Score: 7/10	Criteria	Form A	Form B	Needs More Work	Accom-plished
.1	To estimate, then weigh, using a balance to determine which of two objects is heavier				
.2	To estimate, then weigh, using nonstandard objects				
.3	To estimate, then measure, about how many cups a container will hold				
.4	To classify the temperature of objects as hot or cold				

CHAPTER 22

Score: 8/12	Criteria	Form A	Form B	Needs More Work	Accom-plished
.1	To identify equal parts, halves, fourths, and thirds of a whole				
.2	To visualize results of sharing equal parts to solve problems				
.3	To identify equal parts of a group				

CHAPTER 23

Test Score: 8/12	Criteria	Form A	Form B	Needs More Work	Accom-plished
23-A.1 ☐ ☐	To sort objects and record data in a tally table				
23-A.2 ☐ ☐	To use data to determine if an outcome is certain or impossible, or which event is most likely				
23-A.3 ☐ ☐	To make predictions and record data in tally tables				

CHAPTER 24

Test Score: 8/12	Criteria	Form A	Form B	Needs More Work	Accom-plished
24-A.1 ☐ ☐	To record and interpret data in picture graphs				
24-A.2 ☐ ☐	To count, record, and interpret data in tally tables and bar graphs				

Learning Goals

Child's Name _____ Teacher _____

CHAPTER 25

Test Score:	Criteria	Form A	Form B	Needs More Work	Accom-plished
	9/14	_____	_____		
25-A.1 ☐ ☐	To find sums to 18, using mental math strategies such as *doubles, doubles plus one,* and *doubles minus one*				
25-A.2 ☐ ☐	To write the related addition and subtraction number sentences for doubles fact families				
25-A.3 ☐ ☐	To solve story problems by modeling				

CHAPTER 26

Test Score:	Criteria	Form A	Form B	Needs More Work	Accom-plished
	12/16	_____	_____		
26-A.1 ☐ ☐	To add basic facts with sums 11-18 by using *making a ten and more*				
26-A.2 ☐ ☐	To add three numbers with sums 11-18 by using *doubles* or *making a ten* strategy				
26-A.3 ☐ ☐	To use inverse operations to find sums and differences to 18				

CHAPTER 27

Test Score:	Criteria	Form A	Form B	Needs More Work	Acc plis
	8/12	_____	_____		
27-A.1 ☐ ☐	To make equal groups and count to show how many in all				
27-A.2 ☐ ☐	To put objects into equal groups to determine how many in each group and how many groups				
27-A.3 ☐ ☐	To solve problems by *drawing a picture*				

CHAPTER 28

Test Score:	Criteria	Form A	Form B	Needs More Work	Acc plis
	8/12	_____	_____		
28-A.1 ☐ ☐	To add and subtract tens				
28-A.2 ☐ ☐	To add and subtract tens and ones				
28-A.3 ☐ ☐	To choose a reasonable estimate to solve problems				